This book is excerpts of my notebooks and sketch books of work I have done in the industry with a look into future possibilities.

Special thanks to all who have believed and with determination opened yet another artistic and scientific dimension.

Prefix

Holography was developed by the sharing of both the scientific and artistic communities. Art studios look like science laboratories and in science laboratories scientists are creating artwork. Holographers use both intuitive and subjective aspects of the mind, a discipline well known by Da Vinci, who created with the tools of his time. I became active in holography in the early 1970s, helping to form the first school of holography in Los Angeles, later completing the world's first animated holographic movie in San Francisco. I would like to extend a warm and sincere appreciation to Lloyd Cross and the many other pioneers in the industry for their unselfish sharing of energy in the timeless hours of research; to Posy for her wide exposure of the media to the public; to Herb, Mike, Al and Bob for their sharing and laser arsenal; to Bill and David for their belief; to my family and Kathy who have put up with so many patient hours as the lasers fired, and for God who makes all things possible.

Welcome to the unconquered realm of light.

Holography is the actual recording of the reflected light of objects in 3-D. It is like the light of the sun or lamps that make reality visible, except a regular light is called ambient light. Ambient light is made up of many different frequencies of light waves. Laser means light amplification by stimulated emission of radiation. It is directional light that does not spread out. This light has a radiance power greater than the light emitted from the sun and looks like a projected thread of light.

The frequency of a wave is the number of complete vibrations passing a given point per unit of time. Coherent light is atoms that absorb energy in the light waves, changing from a lower state of energy to one of a higher energy. Millions of atoms emit tiny pulses of light called "photons." Through electrical stimulation to make light directional, light is amplified by reflecting back and forth between two mirrors, one of which is partly a transmission mirror that lets the light out of the active medium.

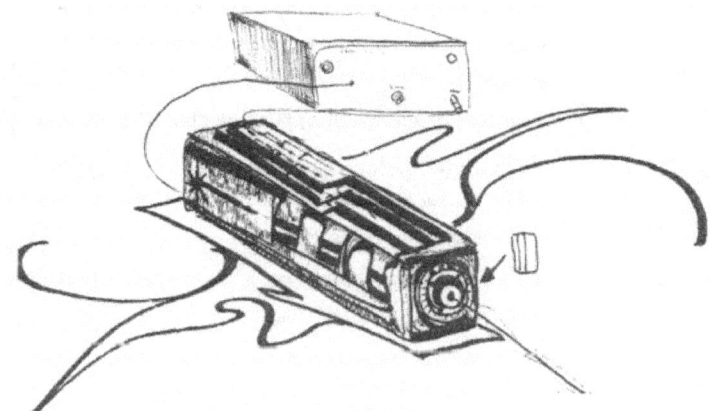

Lasers as a technical tool can transfer 7,000 tv channels, do bloodless surgery, weld an optic nerve, cut through metal, serve as weaponry, increase telephone line capacity -- daily increasing in use where precision is required.

LASER

Gas, helium, krypton and argon lasers are most commonly used for holography. Pulse ruby lasers are used for fast exposures, stopping in a billionth of a second, allowing holograms of projectiles traveling at rapid speeds to be recorded.

A multitude of hyperbolic sets of partial mirrors, created by the two wave fronts interfering upon each other between the reference beam and light from the subject permit the recording of reflected light from all points of the object. These interference patterns were observed by Isaac Newton 300 years ago from distinct wave motion similar to that of intersecting ripples from two pebbles dropped in a pond. Noble prize winner, Dennis Gabor, a scientist from Imperial College of London, discovered phase comparison, viewing not only the brightness, but also spatial relationships of one point to another. Diffraction occurs when a parallel wave front hits on an object and breaks into smaller wave lengths, recording 3-D information on a linear plane.

Holography is a carefully calculated, precise art. For initial experimentation I advise starting with a two milliwatt helium neon laser whose coherence length can be measured by setting up an interferometer, but usually has a one foot coherence length. This is the depth of the scenes that can be recorded because this media is based on recorded light waves. It is imperative to build an isolation table to minimize vibration problems. Any movement of one wave length moving into another during exposure will cancel holograms.

Interference Patterns

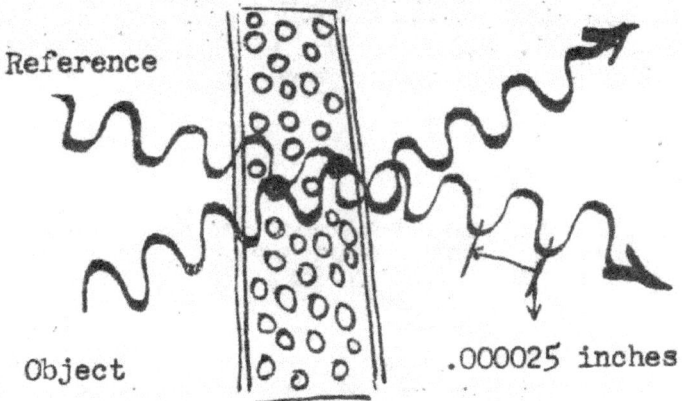

Reference

Object

Emulsion

.000025 inches

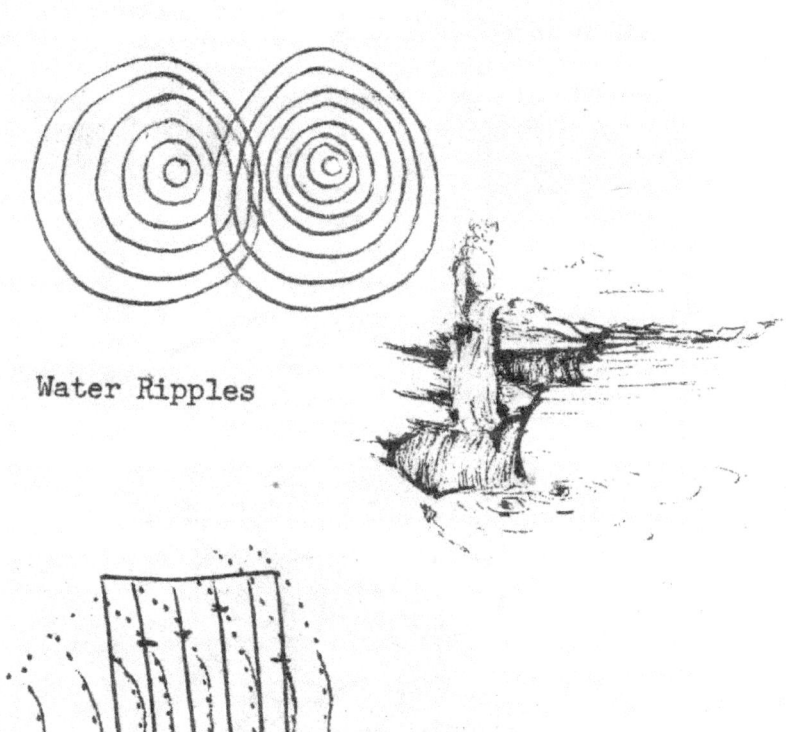

Water Ripples

HONEY COMB METAL TABLE

2, ¼ INCH, 4/4 FOOT STEEL SHEETS. SANDWICHED CANS IN BETWEEN
AND SET ON PLYWOOD PLATFORM WITH INVERTED PYRAMID FOR CENTERING
GRAVITY. SET ON TIRES AND CINDER BLOCKS.

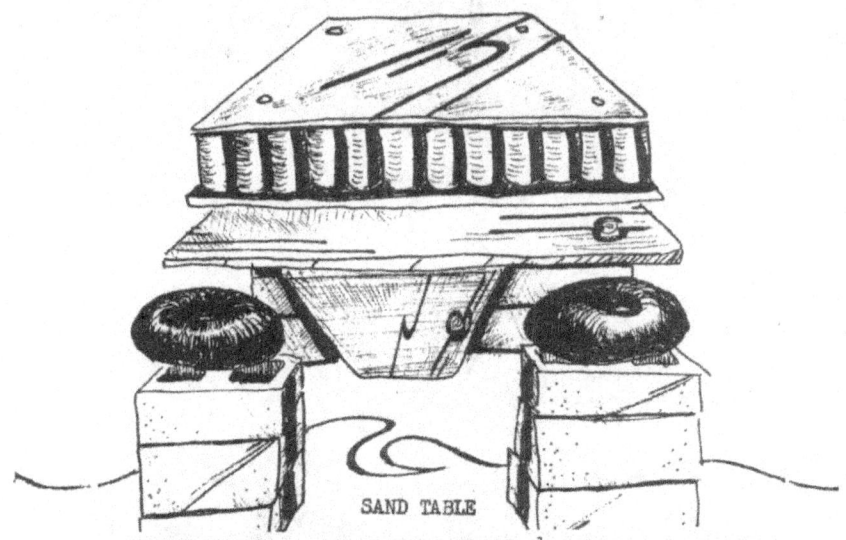

SAND TABLE

CINDER BLOCKS   TIRES AND PLYWOOD, ½ FILLED WITH CEMENT.
THE REST FILLED WITH SILICONE SAND.

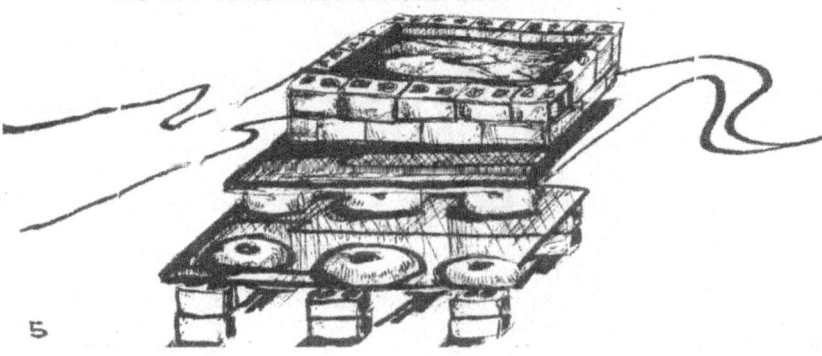

5

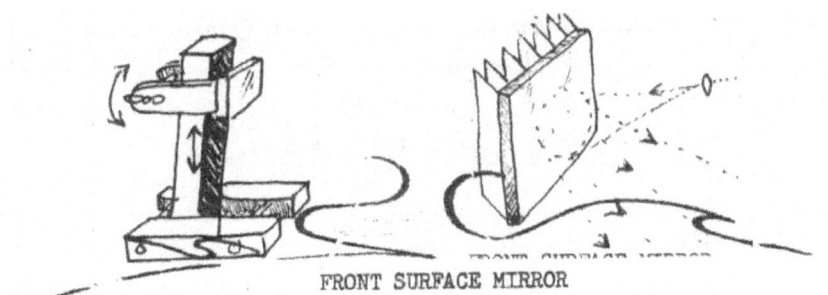

FRONT SURFACE MIRROR

DIFFERENT SIZES ARE NEEDED TO RELAY BEAMS

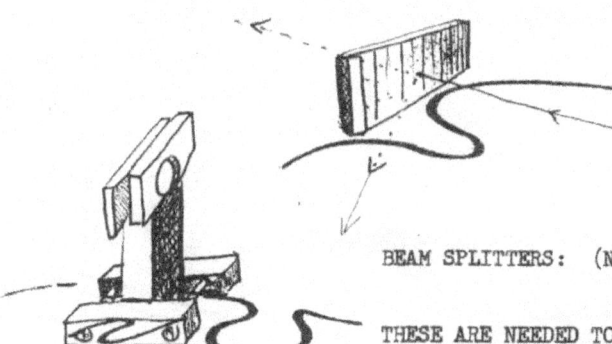

BEAM SPLITTERS: (NEUTRAL DENSITY WEDGE)

THESE ARE NEEDED TO CREATE TWO WAVE FRONTS

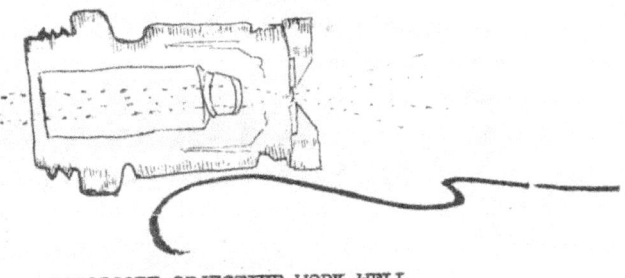

MICROSCOPE OBJECTIVE WORK WELL

FOR EXPANDING BEAMS A PIN HOLE OR SPATIAL
FILTER IS NECESSARY TO HAVE A CLEAN BEAM   6

Interferometry

Used to check table stabilization

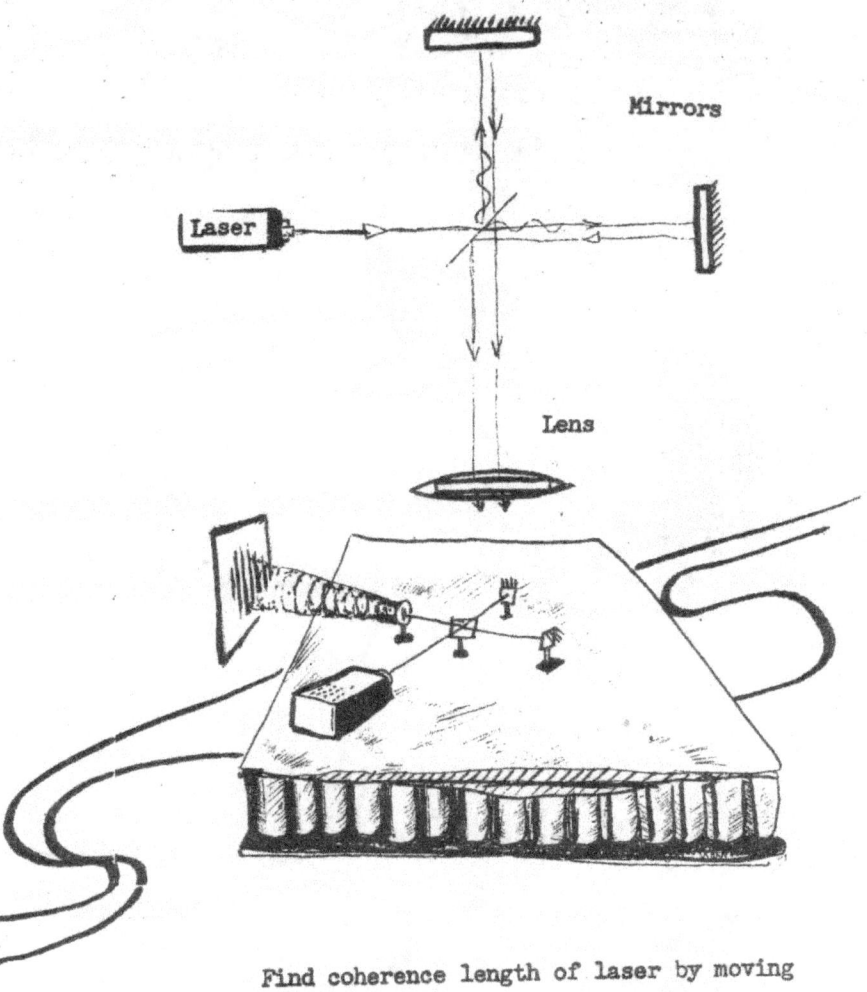

Mirrors

Lens

Find coherence length of laser by moving
mirror apart

FILM HOLDER

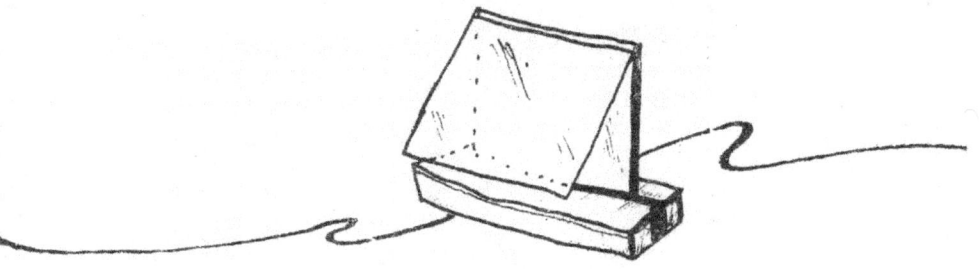

PLATE HOLDER

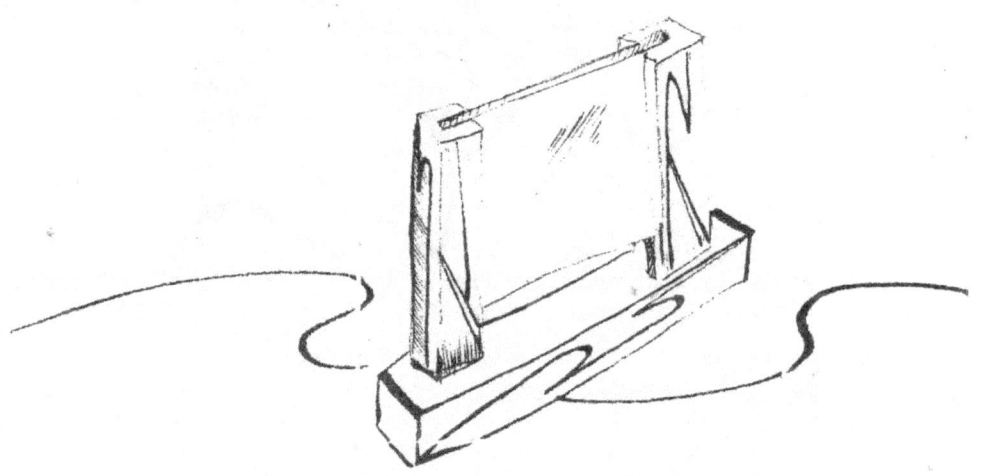

REFLECTION HOLOGRAMS -- MAKING BEST USE OF LOW COHERENCE LASERS.
CO LINEAR LINE HOLOGRAMS CAN BE SEEN IN AMBIENT LIGHT BUT LIMITED
DEPTH. OBJECT MUST BE PLACED NEAR RECORDING PLATE. THIN EMULSION
ORIGINATING FROM LIPPMAN FROM FRANCE WHO USED MERCURY SLIDES FOR
FRONT SURFACE MIRROR OVER 100 YEARS AGO. BECAUSE LASERS WERE NOT
INVENTED UNTIL 1960 HIS WORK ADDED MUCH TO HOLOGRAPHY DEVELOPMENT.
REFLECTION FILM DEVELOPED LATER BY RUSSIAN SCIENTIST, NIFKI.

Object is placed behind photo-sensitive plate
and expanded laser beam goes through film
(reference beam) and bounces back to plate,
thus creating (object beam).

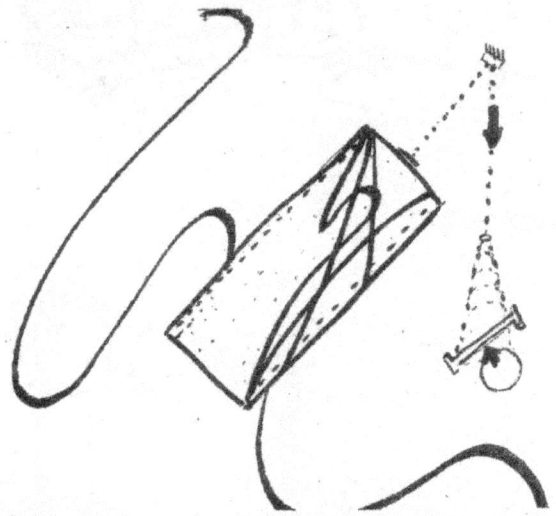

Can be seen with any light source from original reference
beam point on reconstruction but depth is limited.

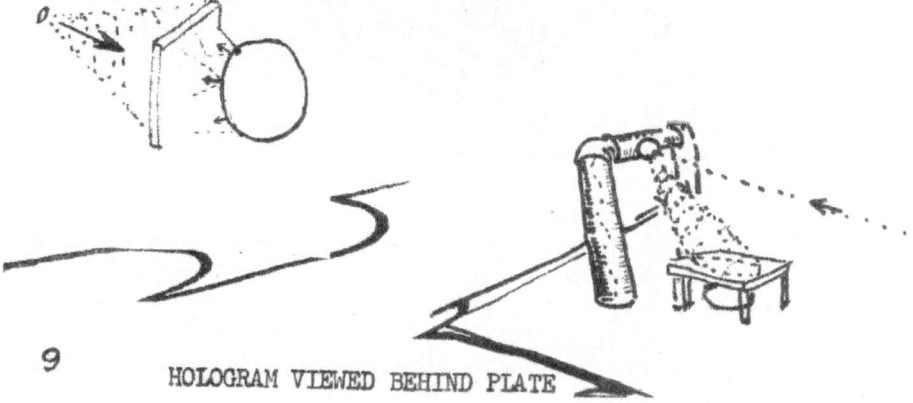

9

HOLOGRAM VIEWED BEHIND PLATE

BEAM PATHS MUST BE EQUAL

Transmission (off-axis hologram) giving
greater depth in scenes developed by E. Leith of
Michigan. Volume hologram must be reconstructed
by same reference angle.

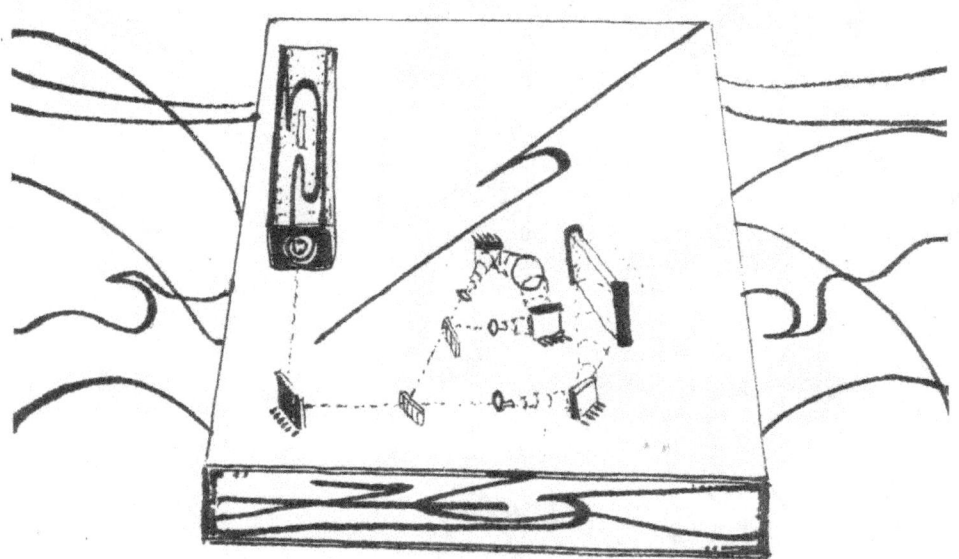

This hologram is formed by object placed in
front of plate. The laser beam is split to create
reference with object beam. Expanded reference
beam is directed at plate. Be sure it doesn't hit
object. Object beam is split again to illuminate
object from two sides.

11

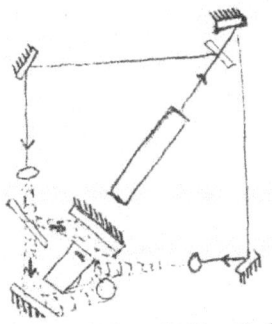

Hologram creates both virtual and real image.
Virtual is through plate and real is object
projected out of plate.

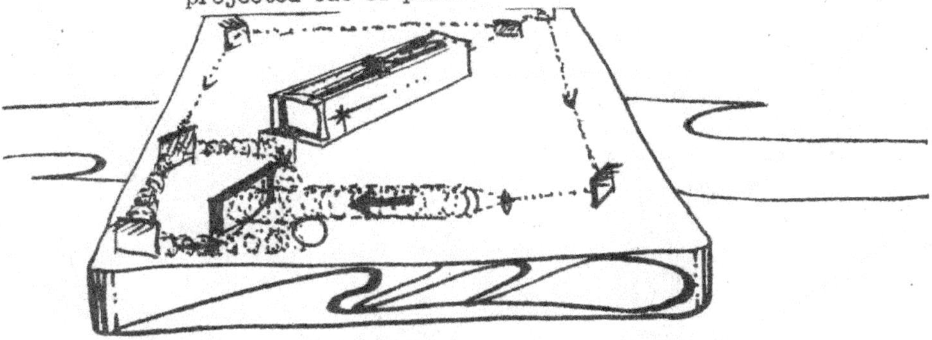

real

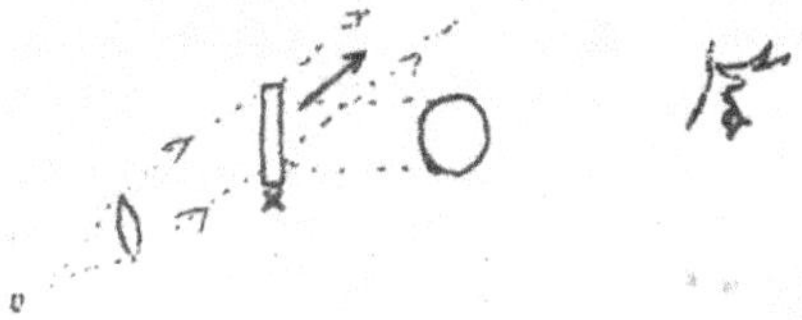

virtual

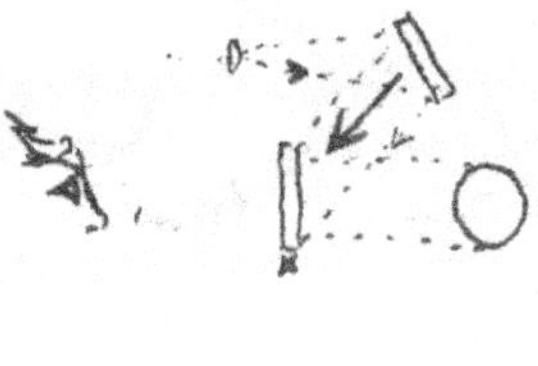

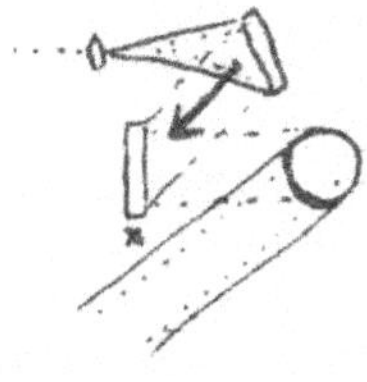

USING PAROBOLIC MIRROR IMAGE PLANE HOLOGRAMS CAN
BE FORMED FOCUSING ON FILM PLATE.

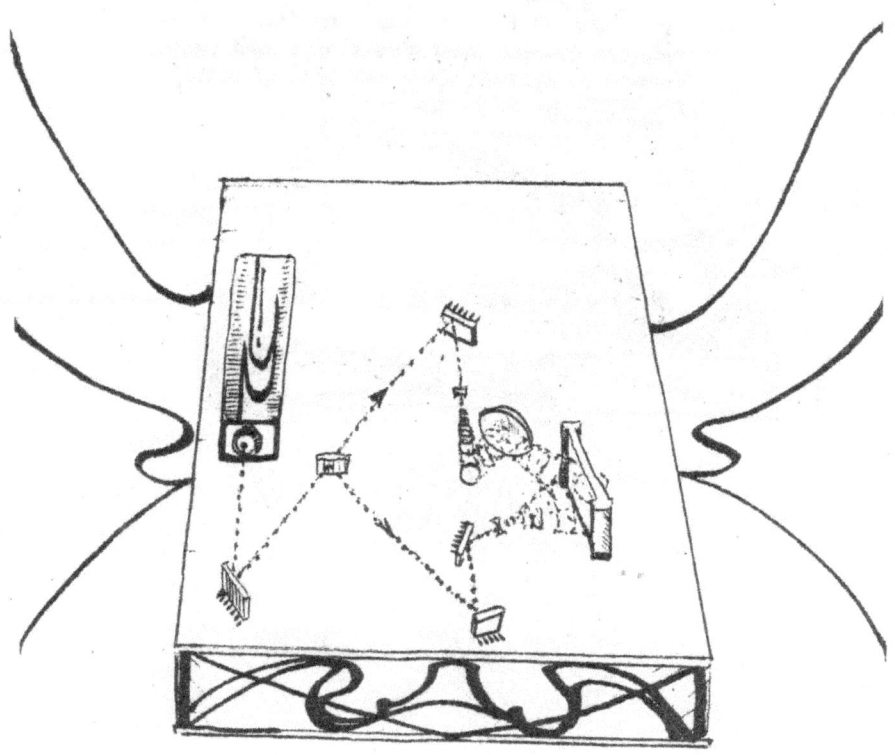

Exposure time = Film sensitivity

$$\frac{\text{Exposure time} = \text{Film sensitivity}}{(\text{Power} \ (\mathcal{M}_k \ ) \ ) \ \times \ 10}$$

A common camera light meter can be used. Bright images must be a one reference to object ratio. Deep or wide images use a possible 1 - 10 ratio.

Processing film -- Transmission

HeNe (8E75AH, 10F 75AH)

1. Develop D-19   5 minutes
2. Stop bath   30 seconds
3. Fix and harden   3 minutes
4. Water wash   10 minutes
5. Bleach in potassium bromide until clear (30 grams), 5 minutes (30 grams potassium ferricyanide)
6. Running wash water

Reflection Hologram (8F75 HAH)

Same, except bleach consists of 20 grams of mercuric chloride and 20 grams of potassium bromide dissolved in 1 liter of water.

Making percent solution

% Solution = $\dfrac{\text{grams of solute}}{\text{total grams of solution}}$

A = grams of solute
B = grams of solvent

$$\% = \frac{A}{A + B}$$

Rainbow holograms are formed from master holograms using cylindrical mirror to fan light through it and bring the reference to copy plate.

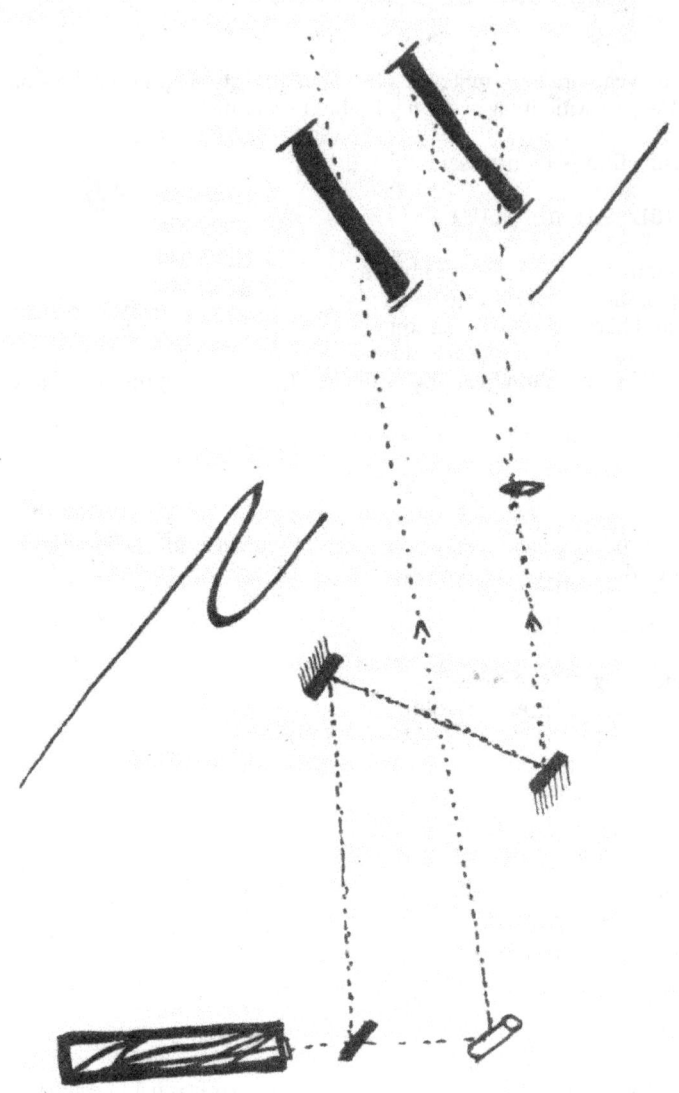

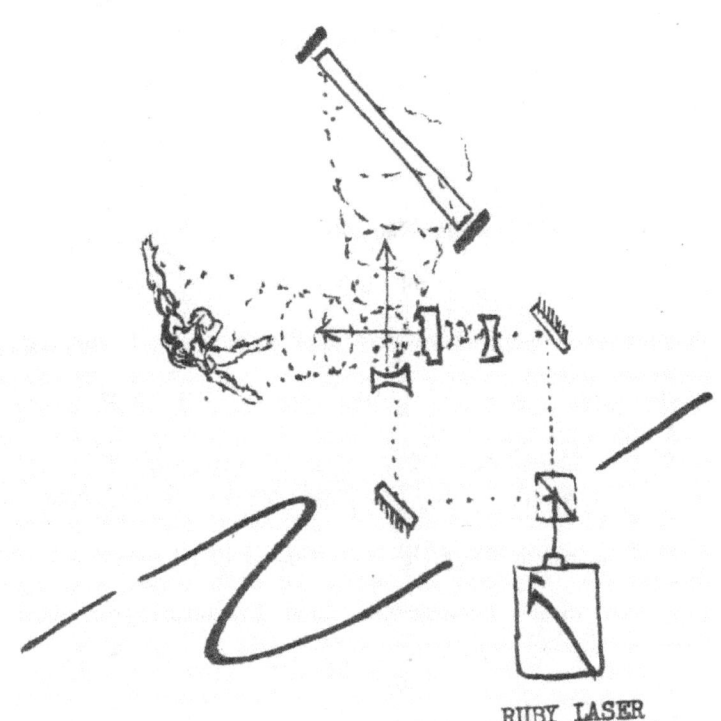

RUBY LASER

Pulse ruby holograms are good for subjects that move because the beam is pulsed at a small fraction of a second and allows some movement in the subject. The ruby laser can shoot holes in metal so good optics must be used to expand beam and a diffusion screen put before the subject. Although, shooting geometries are fairly simple this hologram should be attempted by advanced students. Use slow film such as 649F Kodak plates.

Dichromate holograms use handmade film. One must prepare glass plates with photo sensitive ammonium dichromate and small grade gelatin, 3% dichromate and 10% gelatin. It is best to combine the two together. There are many ways of applying it to glass, but dipping tanks I find work best. Once glass is coated it must dry in the dark and then shot as a reflection hologram with a 1 watt argon laser and 488nm. Developing exposed hologram with water and isopropanol which causes the mass fracturing of the gelatin molecules which cause the hologram to be processed. There is a delicate balance in developing because if the gelatin is fractured too rapidly opalescence occurs; if too slow it will not fracture enough hyperbolic mirrors to have the hologram be bright.

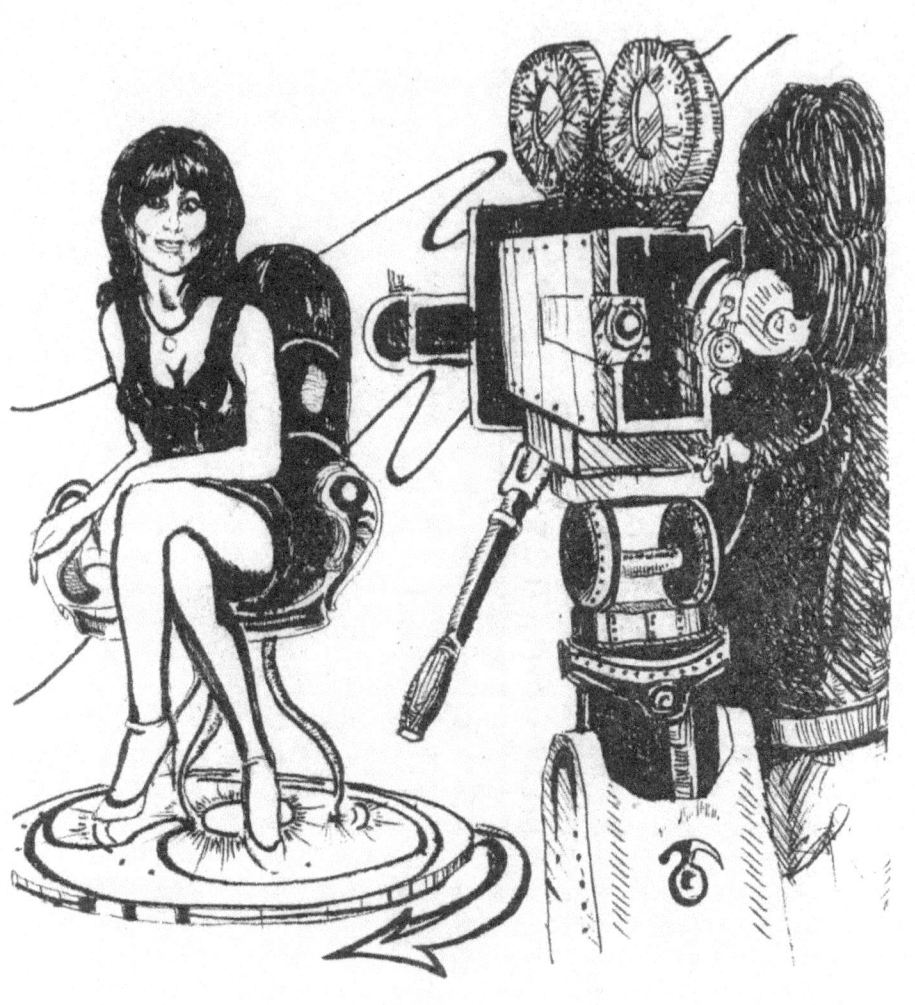

Holographic movies are first filmed by placing the subject on a rotating platform moving at 45 degrees for one complete revolution. Keep movie camera stable. The turn table should be marked in thirds because the printer only transposes film into 3-D holographic film prints only 120 degrees at a time, or 15 seconds (25 feet of positive print black and white film). Use flood lights behind diffusion screens and a bright light directly above subject to give illumination. Also, all movement by your subject must move in slow motion one third real time movement so as not to create time smearing (or blurring) in finished hologram. A number of practice runs before actual shooting is advised. Be sure to take light readings to get correct F stop.

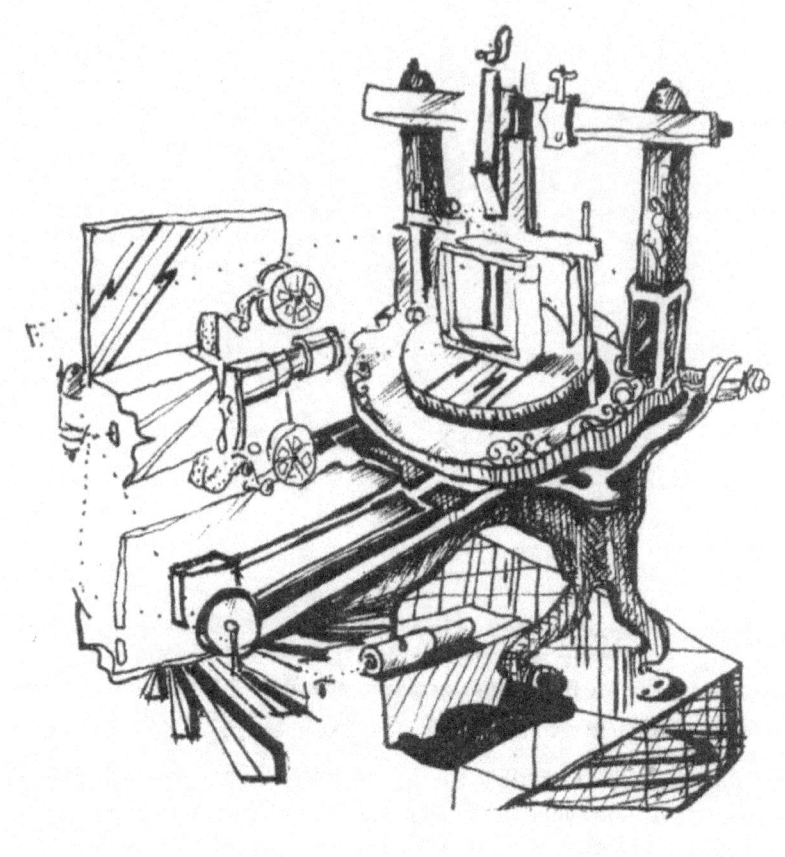

THIS IS A DRAWING OF FIRST MULTIPLEX MARK III
PRINTER THAT PRODUCED THE FIRST HOLOGRAPHIC
MOVIES...DEVELOPED BY LLOYD CROSS.

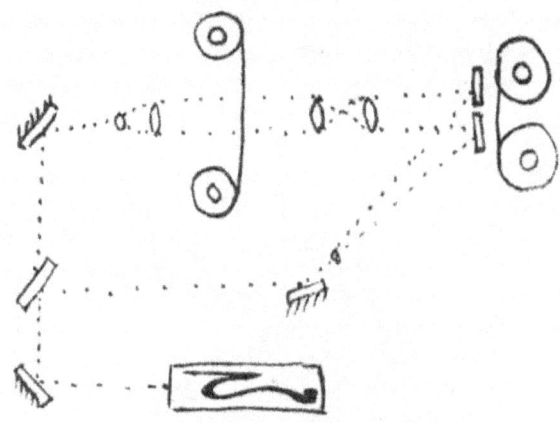

Use a 5 milliwatt HeNe (helium-neon) laser.
Bring object beam through movie film;
then through cylindrical lens to film. Bring
reference from top fan off cylindrical mirror
to film. One to one beam ratio.

Animation drawn on 360 degree parallax.

Scene drawn like it was on a revolving plate. Objects all move in spiral orbits large as they are close, and small as they go into the distance.

A drawing for every third degree of object's movements, four movements every one third of a second, 1080 cells per 360 degrees multiplex holographic movie.

Outdoor filming must be done with three camera positions for every 120 degrees of complete 360 degree holograms. Movement must also be slow motion and movement must move at 45 degree angles to camera. Be sure camera is stable.

Scenes can also be painted behind models and mounted on platform.

This section is from all night continuous discussions I have had with colleagues who with their concern and contributions have given the world yet another dimension with the hopes that the use of the new tool will be for the betterment of our global family.

Entertainment will be holographic with 3-D projections into an audience. This will help break down the barriers of performers and audience. It will be like being there except the actors will be light.

28

People attached to bio feedback mechanisms connected to computers and laser deflection systems can project 3-D images and music. This is possible now.

People can use technology in positive respects to better understand each other by more honest and truthful communication. Individuals will be able to transmit emotions from monitors on their foreheads connected to their nerve center.

Humans with speech defects, autistic or mentally handicapped people could have a better chance of expressing themselves.

Even the possibility of breaking down communication codes of the animal and plant kingdoms is not too remote to consider.

34

There is a great capacity for creativeness through discipline. There is a universe of knowledge existing in our minds and some feel the mind stores information holographically.

People formulate ideas visually and subconsciously.

Our near future travel could likely see the use of our sun in solar powered airships, stabilization in storms by turbines, also giving it direction. It would draw itself to the ground instead of fanning air like a helicopter.

Teaching institutions will use holograms for teaching aids. Libraries will be formed for holographic storage of 3-D reference materials. Medical institutions will teach from holographic body parts. Biologists will have 3-D biological storage. Hologram information storage will be used in all fields. Holographic storage holograms will be used to store book using the space of few filing cabinets instead of taking up a full building like the Library of Congress.

Space travel will use solar ships star chargers with ion bullet reactors for propulsion for covering large areas in space and the possible formation of space colonies.

40

Someday man may travel through light itself, reversing the technology of vaporizing laser weaponry that exists today that explodes matter outward into dust molecules. The technology would be used for transporting instead of destruction by recombining molecules light years away at 186,000 miles per second.

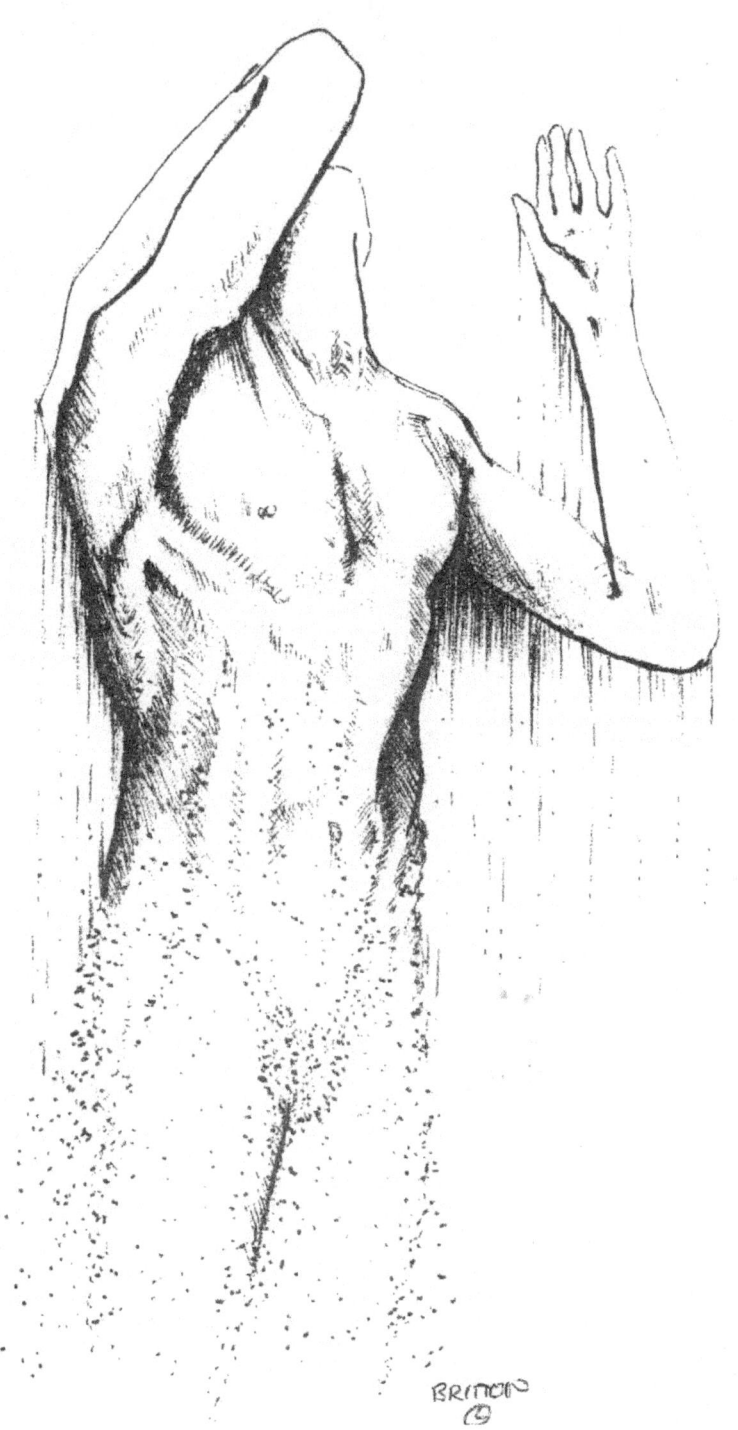

As the doors of the future unfold there is hope people with better communication will harmonize better with nature and use more natural energies such as wind, methane and solar. There is concern that we find a better way to coexist with others in our environment.

I believe further than science or art there is a quest of understanding the actual energy of light as a spiritual entity. As a technological artist my faith is great in our creator for only through him will we all know what's next...